Bees

Everett Franklin Phillips

Alpha Editions

This edition published in 2024

ISBN : 9789367249833

Design and Setting By
Alpha Editions
www.alphaedis.com
Email - info@alphaedis.com

As per information held with us this book is in Public Domain.
This book is a reproduction of an important historical work. Alpha Editions uses the best technology to reproduce historical work in the same manner it was first published to preserve its original nature. Any marks or number seen are left intentionally to preserve its true form.

Contents

INTRODUCTION. ....- 1 -

LOCATION OF THE APIARY. ....- 3 -

EQUIPMENT IN APPARATUS. ....- 6 -

EQUIPMENT IN BEES. ....- 11 -

BEE BEHAVIOR. ....- 15 -

DIRECTIONS FOR GENERAL MANIPULATIONS ....- 21 -

SPRING MANAGEMENT. ....- 29 -

SWARM MANAGEMENT AND INCREASE. ....- 33 -

PREPARATION FOR THE HARVEST. ....- 37 -

THE PRODUCTION OF HONEY. ....- 38 -

THE PRODUCTION OF WAX. ....- 46 -

PREPARATIONS FOR WINTERING. ....- 47 -

DISEASES AND ENEMIES. ....- 50 -

GENERAL INFORMATION. ....- 53 -

INTRODUCTION.

Beekeeping for pleasure and profit is carried on by many thousands of people in all parts of the United States. As a rule, it is not the sole occupation. There are, however, many places where an experienced bee keeper can make a good living by devoting his entire time and attention to this line of work. It is usually unwise to undertake extensive beekeeping without considerable previous experience on a small scale, since there are so many minor details which go to make up success in the work. It is a good plan to begin on a small scale, make the bees pay for themselves and for all additional apparatus, as well as some profit, and gradually to increase as far as the local conditions or the desires of the individual permit.

Bee culture is the means of obtaining for human use a natural product which is abundant in almost all parts of the country, and which would be lost to us were it not for the honey bee. The annual production of honey and wax in the United States makes apiculture a profitable minor industry of the country. From its very nature it can never become one of the leading agricultural pursuits, but that there is abundant opportunity for its growth can not be doubted. Not only is the honey bee valuable as a producer, but it is also one of the most beneficial of insects in cross-pollinating the flowers of various economic plants.

Beekeeping is also extremely fascinating to the majority of people as a pastime, furnishing outdoor exercise as well as intimacy with an insect whose activity has been a subject of absorbing study from the earliest times. It has the advantage of being a recreation which pays its own way and often produces no mean profit.

It is a mistake, however, to paint only the bright side of the picture and leave it to the new bee keeper to discover that there is often another side. Where any financial profit is derived, beekeeping requires hard work and work at just the proper time, otherwise the surplus of honey may be diminished or lost. Few lines of work require more study to insure success. In years when the available nectar is limited, surplus honey is secured only by judicious manipulations, and it is only through considerable experience and often by expensive reverses that the bee keeper is able to manipulate properly to save his crop. Anyone can produce honey in seasons of plenty, but these do not come every year in most locations, and it takes a good bee keeper to make the most of poor years. When, even with the best of manipulations, the crop is a failure through lack of nectar, the bees must be fed to keep them from starvation.

The average annual honey yield per colony for the entire country, under good management, will probably be 25 to 30 pounds of comb honey or 40 to 50 pounds of extracted honey. The money return to be obtained from the crop depends entirely on the market and the method of selling the honey. If sold direct to the consumer, extracted honey brings from 10 to 20 cents per pound, and comb honey from 15 to 25 cents per section. If sold to dealers, the price varies from 6 to 10 cents for extracted honey and from 10 to 15 cents for comb honey. All of these estimates depend largely on the quality and neatness of the product. From the gross return must be deducted from 50 cents to $1 per colony for expenses other than labor, including foundation, sections, occasional new frames and hives, and other incidentals. This estimate of expense does not include the cost of new hives and other apparatus needed in providing for increase in the size of the apiary.

Above all it should be emphasized that the only way to make beekeeping a profitable business is to produce only a first-class article. We can not control what the bees bring to the hive to any great extent, but by proper manipulations we can get them to produce fancy comb honey, or if extracted honey is produced it can be carefully cared for and neatly packed to appeal to the fancy trade. Too many bee keepers, in fact, the majority, pay too little attention to making their goods attractive. They should recognize the fact that of two jars of honey, one in an ordinary fruit jar or tin can with a poorly printed label, and the other in a neat glass jar of artistic design with a pleasing, attractive label, the latter will bring double or more the extra cost of the better package. It is perhaps unfortunate, but nevertheless a fact, that honey sells largely on appearance, and a progressive bee keeper will appeal as strongly as possible to the eye of his customer.

LOCATION OF THE APIARY.

In choosing a section in which to keep bees on an extensive scale it is essential that the resources of the country be known. Beekeeping is more or less profitable in almost all parts of the United States, but it is not profitable to practice extensive beekeeping in localities where the plants do not yield nectar in large quantities. A man who desires to make honey production his business may find that it does not pay to increase the apiaries in his present location. It may be better to move to another part of the country where nectar is more abundant.

FIG. 1.—A well-arranged apiary.

The location of the hives is a matter of considerable importance. As a rule it is better for hives to face away from the prevailing wind and to be protected from high winds. In the North, a south slope is desirable. It is advisable for hives to be so placed that the sun will strike them early in the morning, so that the bees become active early in the day, and thus gain an advantage by getting the first supply of nectar. It is also advantageous to have the hives shaded during the hottest part of the day, so that the bees will not hang out in front of the hive instead of working. They should be so placed that the bees will not prove a nuisance to passers-by or disturb live stock. This latter precaution may save the bee keeper considerable trouble, for bees sometimes prove dangerous, especially to horses. Bees are also sometimes annoying in the early spring, for on their first flights they sometimes spot clothes hung out to dry. This may be remedied by having the apiary some distance from the clothes-drying yard, or by removing the bees from the cellars on days when no clothes are to be hung out.

The plot on which the hives are placed should be kept free from weeds, especially in front of the entrances. The grass may be cut with a lawn mower, but it will often be found more convenient and as efficient to pasture one or more head of sheep in the apiary inclosure.

The hives should be far enough apart to permit of free manipulation. If hives are too close together there is danger of bees entering the wrong hive on returning, especially in the spring.

These conditions, which may be considered as ideal, need not all be followed. When necessary, bees may be kept on housetops, in the back part of city lots, in the woods, or in many other places where the ideal conditions are not found. As a matter of fact, few apiaries are perfectly located; nevertheless, the location should be carefully planned, especially when a large number of colonies are kept primarily for profit.

As a rule, it is not considered best to keep more than 100 colonies in one apiary, and apiaries should be at least 2 miles apart. There are so many factors to be considered, however, that no general rule can be laid down. The only way to learn how many colonies any given locality will sustain is to study the honey flora and the record of that place until the bee keeper can decide for himself the best number to be kept and where they shall be placed.

The experience of a relatively small number of good bee keepers in keeping unusually large apiaries indicates that the capabilities of the average locality are usually underestimated. The determination of the size of extensive apiaries is worthy of considerable study, for it is obviously desirable to keep bees in as few places as possible, to save time in going to them and also expense in duplicated apparatus. To the majority of bee keepers this problem is not important, for most persons keep but a small number of colonies. This is perhaps a misfortune to the industry as a whole, for with fewer apiaries of larger size under the management of careful, trained bee keepers the Honey production of the country would be marvelously increased. For this reason, professional bee keepers are not favorably inclined to the making of thousands of amateurs, who often spoil the location for the honey producer and more often spoil his market by the injudicious selling of honey for less than it is worth or by putting an inferior article on the market.

Out apiaries, or those located away from the main apiary, should be so located that transportation will be as easy as possible. The primary consideration, however, must be the available nectar supply and the number of colonies of bees already near enough to draw on the resources. The out apiary should also be near to some friendly person, so that it may be protected against depredation and so that the owner may be notified if anything goes wrong. It is especially desirable to have it in the partial care of some person who can hive swarms or do other similar things that may arise in an emergency. The terms under which the apiary is placed on land belonging to some one else is a matter for mutual agreement. There is no general usage in this regard.

EQUIPMENT IN APPARATUS.

It can not be insisted too strongly that the only profitable way to keep bees is in hives with movable frames. The bees build their combs in these frames, which can then be manipulated by the bee keeper as necessary. The keeping of bees in boxes, hollow logs, or straw "'skeps'" is not profitable, is often a menace to progressive bee keepers, and should be strongly condemned. Bees in box hives (plain boxes with no frames and with combs built at the will of the bees) are too often seen in all parts of the country. The owners may obtain from them a few pounds of inferior honey a year and carelessly continue in the antiquated practice. In some cases this type of beekeeping does little harm to others, but where diseases of the brood are present the box hive is a serious nuisance and should be abolished.

WORKSHOP.

It is desirable to have a workshop in the apiary where the crop may be cared for and supplies may be prepared. If the ground on which the hives are located is not level, it is usually better to have the shop on the lower side so that the heavier loads will be carried down grade. The windows and doors should be screened to prevent the entrance of bees. The wire-cloth should be placed on the outside of the window frames and should be extended about 6 inches above the opening. This upper border should be held away from the frame with narrow wooden strips one-fourth inch in thickness so as to provide exits for bees which accidentally get into the house. Bees do not enter at such openings, and any bees which are carried into the house fly at once to the windows and then crawl upward, soon clearing the house of all bees. The windows should be so arranged that the glass may be slid entirely away from the openings to prevent bees from being imprisoned. The equipment of benches and racks for tools and supplies can be arranged as is best suited to the house. It is a good plan to provide racks for surplus combs, the combs being hung from strips separated the distance of the inside length of the hive.

HIVES.

It is not the purpose of this bulletin to advocate the use of any particular make of hive or other apparatus. Some general statements may be made, however, which may help the beginner in his choice.

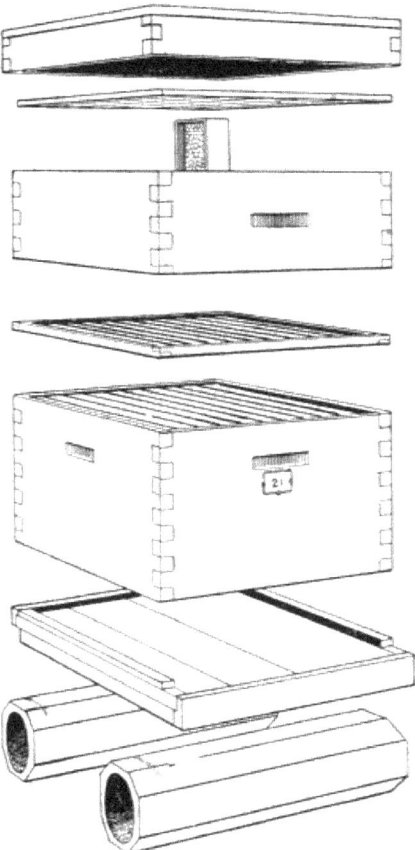

FIG. 2.—A 10-frame hive with comb honey super and perforated zinc queen excluder.

The type of hive most generally used in this country (fig. 2) was invented by Langstroth in 1851. It consists of a plain wooden box holding frames hung from a rabbet at the top and not touching the sides, top, or bottom. Hives of this type are made to hold eight, ten, or more frames. The size of frame in general use, known as the Langstroth (or L) frame ($9^1/_8$ by $17^5/_8$ inches), is more widely used than all others combined. One of the best features in hive manufacture developed by Langstroth is the making of the spaces between frames, side walls, and supers accurately, so that there is just room for the easy passage of bees. In a space of this size (called a "bee space") bees rarely build comb or deposit propolis.

The number of frames used depends on the kind of honey produced (whether comb or extracted) and on the length of honey flow and other local factors. There are other hives used which have points of superiority.

These will be found discussed in the various books on beekeeping and in the catalogues of dealers in bee keepers' supplies. Whatever hive is chosen, there are certain important points which should be insisted on. The material should be of the best; the parts must be accurately made, so that all frames or hives in the apiary are interchangeable. All hives should be of the same style and size; they should be as simple as it is possible to make them, to facilitate operation. Simple frames diminish the amount of propolis, which will interfere with manipulation. As a rule, it is better to buy hives and frames from a manufacturer of such goods rather than to try to make them, unless one is an expert woodworker.

The choice of a hive, while important, is usually given undue prominence in books on bees. In actual practice experienced bee keepers with different sizes and makes of hives under similar conditions do not find as much difference in their honey crop as one would be led to believe from the various published accounts.

Hives should be painted to protect them from the weather. It is usually desirable to use white paint to prevent excessive heat in the colony during hot weather. Other light colors are satisfactory, but it is best to avoid red or black.

FIG. 3.—Smoker.

HIVE STANDS.

Generally it is best to have each hive on a separate stand. The entrance should be lower than any other part of the hive. Stands of wood, bricks, tile (fig. 2), concrete blocks, or any other convenient material will answer the purpose. The hive should be raised above the ground, so that the bottom will not rot. It is usually not necessary to raise the hive more than a few inches. Where ants are a nuisance special hive stands are sometimes necessary.

FIG. 4.—Bee veil with silk-tulle front.

OTHER APPARATUS.

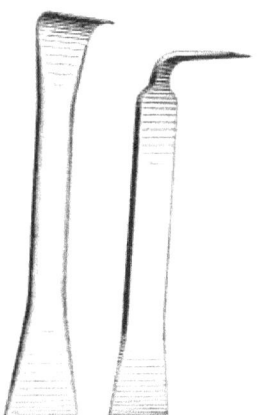

FIG. 5.—Hive tools.

In addition to the hives in which the bees are kept some other apparatus is necessary. A good smoker to quiet the bees (fig. 3), consisting of a tin or copper receptacle to hold burning rotten wood or other material, with a bellows attached, is indispensable. A veil of black material, preferably with a black silk-tulle front (fig. 4), should be used. Black wire-cloth veils are also excellent. Even if a veil is not always used, it is desirable to have one at hand in case the bees become cross. Cloth or leather gloves are sometimes used to protect the hands, but they hinder most manipulations. Some sort of tool (fig. 5) to pry hive covers loose and frames apart is desirable. A screwdriver will answer, but any of the tools made especially for that purpose is perhaps better. Division boards drone traps (fig. 6), bee escapes (figs. 7 and 8), feeders (figs. 17, 18, 19, 20), foundation fasteners, wax extractors, bee brushes (fig. 9), queen-rearing outfits, and apparatus for producing comb or extracted honey (figs. 2, 21, 22) will be found described in catalogues of supplies: a full discussion of these implements would require too much space in this bulletin. A few of these things are illustrated, and their use will be evident to the bee keeper. It is best to have the frames filled with foundation to insure straight combs composed of worker cells only. Foundation is made from thin sheets of pure beeswax on which are impressed the bases of the cells of the comb. On this as a guide the worker bees construct the combs. When sheets of foundation are inserted they should be supported by wires stretched across the frames. Frames purchased from supply dealers are usually pierced for wiring. It should be remembered that manipulation based on a knowledge of bee behavior is of far greater importance than any particular style of apparatus. In a short discussion like the present it is best to omit descriptions of appliances, since supply dealers will be glad to furnish whatever information is desired concerning apparatus.

FIG. 6.—Drone and queen trap on hive entrance.

EQUIPMENT IN BEES.

As stated previously, it is desirable to begin beekeeping with a small number of colonies. In purchasing these it is usually best to obtain them near at home rather than to send to a distance, for there is considerable liability of loss in shipment. Whenever possible it is better to get bees already domiciled in the particular hive chosen by the bee keeper, but if this is not practicable then bees in any hives or in box hives may be purchased and transferred. It is a matter of small importance what race of bees is purchased, for queens of any race may be obtained and introduced in place of the original queen, and in a short time the workers will all be of the same race as the introduced queen. This is due to the fact that during the honey season worker bees die rapidly, and after requeening they are replaced by the offspring of the new queen.

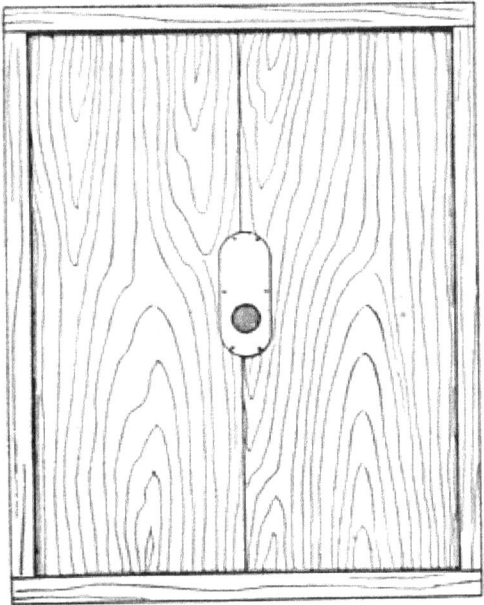

FIG. 7.—Bee escape for removing bees from supers.

A most important consideration in purchasing colonies of bees is to see to it that they are free from disease. In many States and counties there are inspectors of apiaries who can be consulted on this point, but if this is not possible even a novice can tell whether or not there is anything wrong with the brood, and it is always safest to refuse hives containing dead brood.

The best time of the year to begin beekeeping is in the spring, for during the first few months of ownership the bee keeper can study the subject and learn what to do, so that he is not so likely to make a mistake which will end in loss of bees. It is usually best to buy good strong colonies with plenty of brood for that season of the year, but if this is not practicable, then smaller colonies, or nuclei, may be purchased and built up during the summer season. Of course, no surplus honey can be expected if all the honey gathered goes into the making of additional bees. It is desirable to get as little drone comb as possible and a good supply of honey in the colonies purchased.

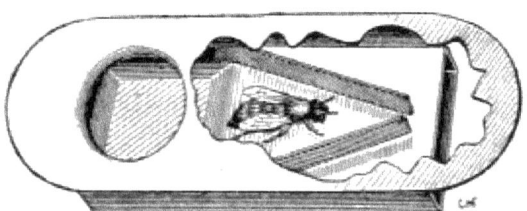

FIG. 8.—Spring bee escape.

The question as to what race and strain of bees is to be kept is important. If poor stock has been purchased locally, the bee keeper should send to some reliable queen breeder for good queens as a foundation for his apiary. Queens may be purchased for $1 each for "untested" to several dollars each for "selected" breeding queens. Usually it will not pay beginners to buy "selected" breeding queens, for they are not yet prepared to make the best use of such stock. "Untested" or "tested" queens are usually as good a quality as are profitable for a year or so, and there is also less danger in mailing "untested" (young) queens.

Various races of bees have been imported into the United States and among experienced bee keepers there are ardent advocates of almost all of them. The black or German race was the first imported, very early in the history of the country, and is found everywhere, but usually not entirely pure. As a rule this race is not desirable. No attention has been paid to breeding it for improvement in this country, and it is usually found in the hands of careless bee keepers. As a result it is inferior, although it often produces beautiful comb honey.

The Italian bees, the next introduced, are the most popular race among the best bee keepers in this country, and with good reason. They are vigorous workers and good honey gatherers, defend their hives well, and above all have been more carefully selected by American breeders than any other race. Especially for the last reason it is usually desirable to keep this race. That almost any other race of bees known could be bred to as high a point as the Italians, and perhaps higher, can not be doubted, but the bee keeper now gets the benefit of what has been done for this race. It should not be understood from this that the efforts at breeding have been highly successful. On the contrary, bee breeding will compare very unfavorably with the improvement of other animals or plants which have been the subject of breeding investigations.

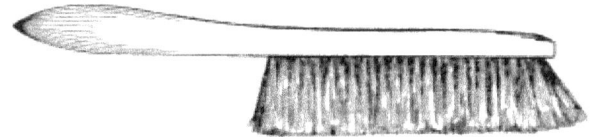

FIG. 9.—Bee brush.

Italian bees have been carefully selected for color by some breeders to increase the area of yellow on the abdomen, until we now have what are known as "five-banded" bees. These are very beautiful, but it can scarcely be claimed that they are improved as honey producers or in regard to gentleness. They are kept mostly by amateurs. Some breeders have claimed to select Italians for greater length of tongue, with the object of getting a bee which could obtain the abundance of nectar from red clover. If any gain is ever made in this respect, it is soon lost. The terms "red-clover bees" or "long-tongued bees" are somewhat misleading, but are ordinarily used as indicating good honey producers.

Caucasian bees, formerly distributed throughout the country by this department, are the most gentle race of bees known. They are not stingless, however, as is often stated in newspapers and other periodicals. Many report them as good honey gatherers. They are more prolific than Italians and may possibly become popular. Their worst characteristic is that they gather great quantities of propolis and build burr and brace combs very freely. They are most desirable bees for the amateur or for experimental purposes.

Carniolan and Banat bees have some advocates, and are desirable in that they are gentle. Little is known of Banats in this country, Carniolans swarm excessively unless in' large hives. Cyprians were formerly used somewhat, but are now rarely found pure, and are undesirable either pure or in crosses because of the fact that they sting with the least provocation and are not manageable with smoke. They are good honey gatherers, but their undesirable qualities have caused them to be discarded by American bee keepers. "Holy-land," Egyptian, and Punic (Tunisian) bees have also been tried and have been universally abandoned.

The Department of Agriculture does not now distribute or sell queen bees or colonies of bees of any race.

BEE BEHAVIOR.

The successful manipulation of bees depends entirely on a knowledge of their habits. This is not generally recognized, and most of the literature on practical beekeeping consists of sets of rules to guide manipulations. This is too true of the present paper, but is due to a desire to make the bulletin short and concise. While this method usually answers, it is nevertheless faulty, in that, without a knowledge of fundamental principles of behavior, the bee keeper is unable to recognize the seemingly abnormal phases of activity, and does not know what to do under such circumstances. Rules must, of course, be based on the usual behavior. By years of association the bee keeper almost unconsciously acquires a wide knowledge of bee behavior, and consequently is better able to solve the problems which constantly arise. However, it would save an infinite number of mistakes and would add greatly to the interest of the work if more time were expended on a study of behavior; then the knowledge gained could be applied to practical manipulation.

A colony of bees consists normally of one queen bee (fig. 10, b), the mother of the colony, and thousands of sexually undeveloped females called workers (fig. 10, a), which normally lay no eggs, but build the comb, gather the stores, keep the hive clean, feed the young, and do the other work of the hive. During part of the year there are also present some hundreds of males (fig. 10, c) or drones (often removed or restricted in numbers by the bee keeper), whose only service is to mate with young queens. These three types are easily recognized, even by a novice. In nature the colony lives in a hollow tree or other cavity, but under manipulation thrives in the artificial hives provided. The combs which form their abode are composed of wax secreted by the workers. The hexagonal cells of the two vertical layers constituting each comb have interplaced ends on a common septum. In the cells of these combs are reared the developing bees, and honey and pollen for food are also stored here.

The cells built naturally are not all of the same size, those used in rearing worker bees being about one-fifth of an inch across, and those used in rearing drones and in storing honey about one-fourth of an inch across (fig. 11). The upper cells in natural combs are more irregular, and generally curve upward at the outer end. They are used chiefly for the storage of honey. Under manipulation the size of the cells is controlled by the bee keeper by the use of comb foundation—sheets of pure beeswax on which are impressed the bases of cells and on which the bees build the side walls.

FIG. 10.—The honey bee: a, Worker; b, queen; c, drone. Twice natural size.

In the North, when the activity of the spring begins, the normal colony consists of the queen and some thousands of workers. As the outside temperature raises, the queen begins to lay eggs (fig. 12, a) in the worker cells. These in time develop into white larvæ (fig. 12, b, c), which grow to fill the cells. They are then capped over and transform first into pupæ (fig. 12, d) and then into adult worker bees. As the weather grows warmer, and the colony increases in size by the emergence of the young bees, the quantity of brood is increased. The workers continue to bring in pollen, nectar to be made into honey, and water for brood rearing. When the hive is nearly filled with bees and stores, or when a heavy honey flow is on, the queen begins to lay eggs in the larger cells, and these develop into drones or males. Continued increase of the colony would result in the formation of enormous, colonies, and unless some division takes place no increase in the number of colonies will result. Finally, however, the workers begin to build queen cells (fig. 13). These are larger than any other cells In the hive and hang on the comb vertically. In size and shape they may be likened to a peanut, and are also rough on the outside. In preparing for swarming the queen sometimes lays eggs in partly constructed queen cells, but when a colony becomes queenless the cells are built around female larvæ. The larvæ in these cells receive special food, and when they have grown to full size they, too, are sealed up, and the colony is then ready for swarming.

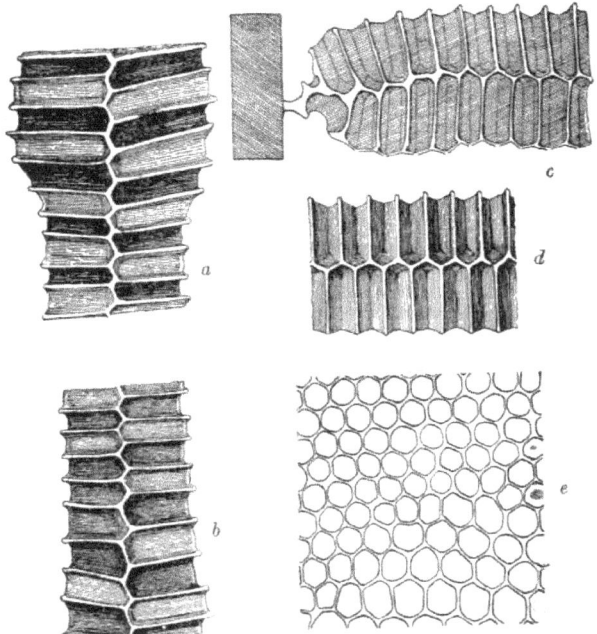

FIG. 11.—Comb architecture: *a*, Vertical section at top of comb; *b*, vertical section showing transition from worker to drone cells; *c*, horizontal section at side of comb showing end bar of frame; *d*, horizontal section of worker brood cells; *e*, diagram showing transition cells. Natural size.

The issuing of the first swarm from a colony consists of the departure of the original queen with part of the workers. They leave behind the Honey stores, except such as they can carry in their honey stomachs, the brood, some workers, drones, several queen cells, from which will later emerge young queens, but no adult queen. By this interesting process the original colony is divided into two.

The swarm finds a new location in some place, such as a hollow tree, or, if cared for by the bee keeper, in a hive. The workers build new combs, the queen begins laying, and in a short time the swarm becomes a normal colony.

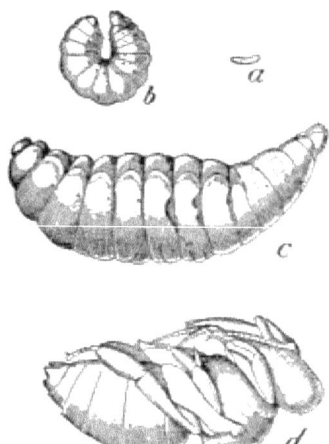

FIG. 12.—The honey bee: *a*, Egg; *b*, young larva; *c*, old larva; *d*, pupa. Three times natural size.

The colony on the old stand (parent colony) is increased by the bees emerging from the brood. After a time (usually about seven or eight days) the queens in their cells are ready to emerge. If the colony is only moderately strong the first queen to emerge is allowed by the workers to tear down the other queen cells and kill the queens not yet emerged, but if a "second swarm" is to be given off the queen cells are protected.

If the weather permits, when from 5 to 8 days old, the young queen flies from the hive to mate with a drone. Mating usually occurs but once during the life of the queen and always takes place on the wing. In mating she receives enough spermatozoa (male sex cells) to last throughout her life. She returns to the hive after mating, and in about two days begins egg laying. The queen never leaves the hive except at mating time or with a swarm, and her sole duty in the colony is to lay eggs to keep up the population.

When the flowers which furnish most nectar are in bloom, the bees usually gather more honey than they need for their own use, and this the bee keeper can safely remove. They continue the collection of honey and other activities until cold weather comes on in the fall, when brood rearing ceases; they then become relatively quiet, remaining in the hive all winter, except for short flights on warm days. When the main honey flow is over, the drones are usually driven from the hive. By that time the virgin queens have been mated and drones are of no further use. They are not usually stung to death, but are merely carried or driven from the hive by the workers and starve. A colony of bees which for any reason is without a queen does not expel the drones.

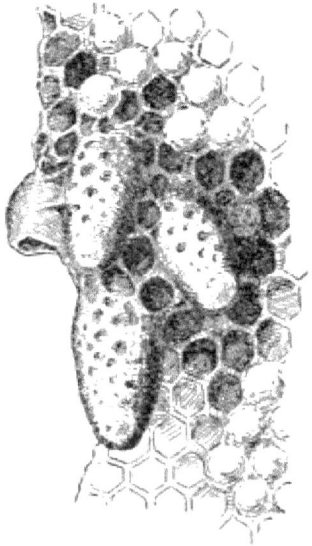

FIG. 13.—Queen cells. Natural size.

Many abnormal conditions may arise in the activity of a colony, and it is therefore necessary for the bee keeper to understand most of these, so that when they occur he may overcome them. If a virgin queen is prevented from mating she generally dies, but occasionally begins to lay eggs after about four weeks. In this event, however, all of the eggs which develop become males. Such a queen is commonly called a "drone layer."

If the virgin queen is lost while on her flight, or the colony at any other time is left queenless without means of rearing additional queens, it sometimes happens that some of the workers begin to lay eggs. These eggs also develop only into drones.

It also happens at times that when a queen becomes old her supply of spermatozoa is exhausted, at which tune her eggs also develop only into drones. These facts are the basis of the theory that the drone of the bee is developed from an unfertilized egg or is partheno-genetic. A full discussion of this point is impossible in this place.

The work of the hive is very nicely apportioned among the inmates, so that there is little lost effort. As has been stated, the rearing of young is accomplished by having one individual to lay eggs and numerous others (immature females or workers) to care for the larvæ. In like manner all work of the colony is apportioned. In general, it may be stated that all inside work—wax building, care of brood, and cleaning—is done by the younger workers, those less than 17 days old, while the outside work of collecting pollen and nectar to be made into honey is done by the older workers. This plan may be changed by special conditions. For example, if the colony has been queenless for a tune and a queen is then given, old workers may begin the inside work of feeding larvæ, and these may also secrete wax. Or, if the old workers are all removed, the younger bees may begin outside work. As a rule, however, the general plan of division of labor according to age is probably followed rather closely.

DIRECTIONS FOR GENERAL MANIPULATIONS.

Bees should be handled so that they will be little disturbed in their work. As much as possible, stings should be avoided during manipulation. This is true, not so much because they are painful to the operator, but because the odor of poison which gets into the air irritates the other bees and makes them more difficult to manage. For this reason it is most advisable to wear a black veil (fig. 4) over a wide-brimmed hat and to have a good smoker (fig. 3). Gloves, however, are usually more an inconvenience than otherwise. Gauntlets or rubber bands around the cuffs keep the bees from crawling up the sleeve. It is best to avoid black clothing, since that color seems to excite bees; a black felt hat is especially to be avoided.

Superfluous quick movements tend to irritate the bees. The hive should not be jarred or disturbed any more than necessary. Rapid movements are objectionable, because with their peculiar eye structure bees probably perceive motion more readily than they do objects. Persons not accustomed to bees, on approaching a hive, often strike at bees which fly toward them or make some quick movement of the head or hand to avoid the sting which they fear is to follow. This should not be done, for the rapid movement, even if not toward the bee, is far more likely to be followed by a sting than remaining quiet.

The best time to handle bees is during the middle of warm days, particularly during a honey flow. Never handle bees at night or on cold, wet days unless absolutely necessary. The work of a beginner may be made much easier and more pleasant by keeping gentle bees. Caucasians, Carniolans, Banats, and some strains of Italians ordinarily do not sting much unless unusually provoked or except in bad weather. Common black bees or crosses of blacks with other races are more irritable. It may be well worth while for the beginner to procure gentle bees while gaining experience in manipulation. Later on, this is less important, for the bee keeper learns to handle bees with little inconvenience to himself or to the bees. Various remedies for bee stings have been advocated, but they are all useless. The puncture made by the sting is so small that it closes when the sting is removed and liquids can not be expected to enter. The best thing to do when stung is to remove the sting as soon as possible without squeezing the poison sac, which is usually attached. This can be done by scraping it out with a knife or finger nail. After this is done the injured spot should be let alone and not rubbed with any liniment. The intense itching will soon disappear; any irritation only serves to increase the afterswelling.

Before opening a hive the smoker should be lighted and the veil put on. A few puffs of smoke directed into the entrance will cause the bees to fill themselves with honey and will drive back the guards. The hive cover should be raised gently, if necessary being pried loose with a screwdriver or special hive tool. When slightly raised, a little more smoke should be blown in vigorously on the tops of the frames, or if a mat covering for the frames is used, the cover should be entirely removed and one corner of the mat lifted to admit smoke. It is not desirable to use any more smoke than just enough to subdue the bees and keep them down on the frames. If at any time during manipulation they become excited, more smoke may be necessary. Do not stand in front of the entrance, but at one side or the back.

After the frames are exposed they may be loosened by prying gently with the hive tool and crowded together a little so as to give room for the removal of one frame. In cool weather the propolis (bee glue) may be brittle. Care should be exercised not to loosen this propolis with a jar. The first frame removed can be leaned against the hive, so that there will be more room inside for handling the others. During all manipulations bees must not be mashed or crowded, for it irritates the colony greatly and may make it necessary to discontinue operations. Undue crowding may also crush the queen. If bees crawl on the hands, they may be gently brushed off or thrown off.

FIG. 14.—Handling the frame: First position.

In examining a frame hold it over the hive if possible, so that any bees or queen which fall may drop into it. Freshly gathered honey also often drops from the frame, and if it falls in the hive the bees can quickly clean it up, whereas if it drops outside it is untidy and may cause robbing. If a frame is temporarily leaned against the hive, it should be placed in a nearly upright position to prevent breakage and leaking of honey. The frame on which the queen is located should not be placed on the ground, for fear she may crawl away and be lost. It is best to lean the frame on the side of the hive away from the operator, so that bees will not crawl up his legs.

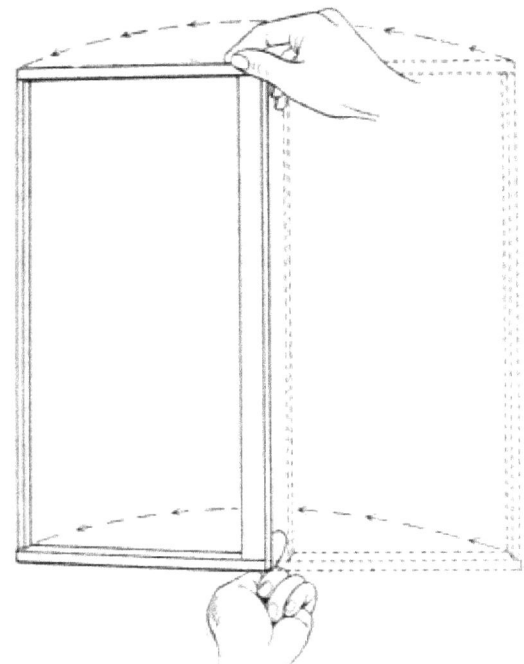

FIG. 15.—Handling the frame: Second position.

In hanging frames the comb should always be held in a vertical position, especially if it contains much honey. When a frame is lifted from the hive by the top bar, the comb is vertical with one side toward the operator (fig. 14). To examine the reverse side, raise one end of the top bar until it is perpendicular (fig. 15), turn the frame on the top bar as an axis until the reverse side is in view, and then lower to a horizontal position with the top bar below (fig. 16). In this way there is no extra strain on the comb and the bees are not irritated. This care is not so necessary with wired combs, but it is a good habit to form in handling frames.

It is desirable to have combs composed entirely of worker cells in order to reduce the amount of drone brood. The use of full sheets of foundation will bring this about and is also of value in making the combs straight, so that bees are not mashed in removing the frame. It is extremely difficult to remove combs built crosswise in the hive, and this should never be allowed to occur. Such a hive is even worse than a plain box hive. Superfluous inside fixtures should be avoided, as they tend only to impede manipulation. The hive should also be placed so that the entrance is perfectly horizontal and a little lower than the back of the hive. The frames will then hang in a vertical position, and the outer ones will not be fastened by the bees to the hive body if properly spaced at the top.

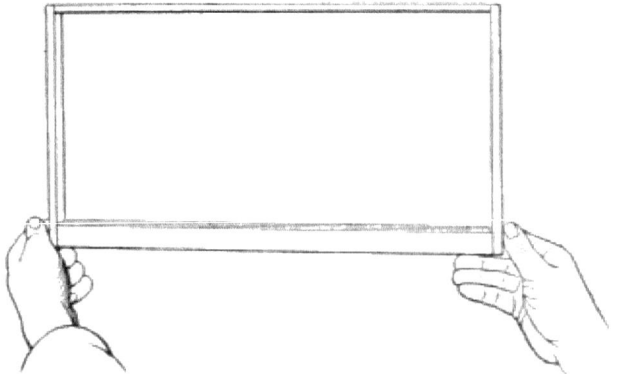
FIG. 10.—Handling the frame: Third position.

In placing frames in the hive great care should be exercised that they are properly spaced. Some frames are self-spacing, having projections on the side, so that when placed as close as possible they are the correct distance apart. These are good for beginners or persons who do not judge distances well and are preferred by many professional bee keepers. If unspaced frames are used, the brood frames should be $1^3/_8$ inches from center to center. A little practice will usually enable anyone to space quickly and accurately. Careful spacing is necessary to prevent the building of combs of irregular thickness and to retard the building of pieces of comb from one frame to another.

A beginner in beekeeping should by all means, if possible, visit some experienced bee keeper to get suggestions in handling bees. More can be learned in a short visit than in a considerably longer time in reading directions, and numerous short cuts which are acquired by experience will well repay the trouble or expense of such a visit. Not all professional bee keepers manipulate in the very best way, but later personal experience will correct any erroneous information. Above all, personal experimentation and a study of bee activity are absolute necessities in the practical handling of bees.

TRANSFERRING.

In increasing the apiary it is sometimes best to buy colonies in box hives on account of their smaller cost and to transfer them to hives with movable frames. This should be done as soon as possible, for box hive colonies are of small value as producers. The best time to transfer is in the spring (during fruit bloom in the North) when the amount of honey and the population of the colony are at a minimum.

Transferring should not be delayed until spring merely because that season is best for the work. It may be done at any time during the active season, but, whenever possible, during a honey flow, to prevent robbing. If necessary, it may be done in a tent such as is often used in manipulating colonies. By choosing a time of the day when the largest number of bees are in the field the work will be lessened.

Plan 1.—The box hive should be moved a few feet from its stand and in its place should be put a hive with movable frames containing full sheets of foundation. The box hive should be turned upside down and a small, empty box inverted over it. By drumming continuously on the box hive with sticks for a considerable time the bees will be made to desert their combs and go to the upper box, and when most of them are clustered above, the bees may be dumped in front of the entrance of the hive which is to house them. The queen will usually be seen as the bees enter the hive, but, in case she has not left the old combs, more drumming will induce her to do so. It is necessary that the queen be in the hive before this manipulation is finished. The old box hive containing brood may now be placed right side up in a new location and in 21 days all of the worker brood will have emerged and probably some new queens will have been reared. These bees may then be drummed out and united with their former hive mates by vigorously smoking the colony and the drummed bees and allowing the latter to enter the hive through a perforated zinc to keep out the young queens. The comb in the box hive may then be melted up and any honey which it may contain used as the bee keeper sees fit. By this method good straight combs are obtained. If little honey is being gathered, the colony in the hive must be provided with food.

Plan 2.—If, on the other hand, the operator desires to save the combs of the box hive, the bees may be drummed into a box and the brood combs and other fairly good combs cut to fit frames and tied in place or held with rubber bands, strings, or strips of wood until the bees can repair the damage and fill up the breaks. These frames can then be hung in a hive on the old stand and the bees allowed to go in. The cutting of combs containing brood with more or less bees on them is a disagreeable job, and, since the combs so obtained are usually of little value in an apiary, the first method is recommended.

Plan 3.—Another good plan is to wait until the colony swarms and then move the box hive to one side. A movable frame hive is now placed in the former location of the box hive and the swarm is hived in it. In this way all returning field bees are forced to join the swarm. In 21 days all of the worker brood in the box hive will have emerged. These young bees may then be united with the bees in the frame hive and the box hive destroyed.

Colonies often take up their abode in walls of houses and it is often necessary to remove them to prevent damage from melting combs. If the cavity in which the combs are built can be reached, the method of procedure is like that of transferring, except that drumming is impractical and the bees must simply be subdued with smoke and the combs cut out with the bees on them.

Another method which is often better is to place a bee escape over the entrance to the cavity, so that the bees can come out, but can not return. A cone of wire cloth about 8 inches high with a hole at the apex just large enough for one bee to pass will serve as a bee escape, or regular bee escapes (fig. 8) such as are sold by dealers may be used. A hive which they can enter is then placed beside the entrance. The queen is not obtained in this way and, of course, goes right on laying eggs, but as the colony is rapidly reduced in size the amount of brood decreases. As brood emerges, the younger bees leave the cavity and join the bees in the hive, until finally the queen is left practically alone. A new queen should be given to the bees in the hive as soon as possible, and in a short time they are fully established in their new quarters. After about four weeks, when all or nearly all of the brood in the cavity has emerged, the bee escape should be removed and as large a hole made at the entrance of the cavity as possible. The bees will then go in and rob out the honey and carry it to the hive, leaving only empty combs. The empty combs will probably do no damage, as moths usually soon destroy them and they may be left in the cavity and the old entrance carefully closed to prevent another swarm from taking up quarters there.

In transferring bees from a hollow tree the method will depend on the accessibility of the cavity. Usually it is difficult to drum out the bees and the combs can be cut out after subduing the colony with smoke.

UNITING.

Frequently colonies become queenless when it is not practicable to give them a new queen, and the best practice under such conditions is to unite the queenless bees to a normal colony. If any colonies are weak in the fall, even if they have a queen, safe wintering is better insured if two or more weak colonies are united, keeping the best queen. Under various other conditions which may arise the bee keeper may find it desirable to unite bees from different colonies. Some fundamental facts in bee behavior must be thoroughly understood to make this a success.

Every colony of bees has a distinctive colony odor and by this means bees recognize the entering of their hive by bees from other colonies and usually resent it. If, however, a bee comes heavily laden from the field and flies directly into the wrong hive without hesitation it is rarely molested. In uniting colonies, the separate colony odors must be hidden, and this is done by smoking each colony vigorously. It may at times be desirable to use tobacco smoke, which not only covers the colony odor but stupefies the bees somewhat. Care should be taken not to use too much tobacco, as it will completely overcome the bees. The queen to be saved should be caged for a day or two to prevent the strange bees from killing her in the first excitement.

Another fact which must be considered is that the bees of a colony carefully mark the location of their own hive and remember that location for some time after they are removed. If, therefore, two colonies in the apiary which are not close together are to be united, they should be moved gradually nearer, not more than a foot at a time, until they are side by side, so that the bees will not return to their original locations and be lost. As the hives are moved gradually the slight changes are noted and no such loss occurs. As a further precaution, a board should be placed in front of the entrance in a slanting position, or brush and weeds may be thrown down so that when the bees fly out they recognize the fact that there has been a change and accustom themselves to the new place. If uniting can be done during a honey flow, there is less danger of loss of bees by fighting, or if done in cool weather, when the bees are not actively rearing brood, the colony odors are diminished and the danger is reduced.

It is an easy matter to unite two or more weak swarms to make one strong one, for during swarming the bees have lost their memory of the old location, are full of honey, and are easily placed wherever the bee keeper wishes. They may simply be thrown together in front of a hive. Swarms may also be given to a newly established colony with little difficulty.

PREVENTING ROBBING IN THE APIARY.

When there is no honey flow bees are inclined to rob other colonies, and every precaution must be taken to prevent this. Feeding often attracts other bees, and, if there are indications of robbing, the sirup or honey should be given late in the day. As soon as robbing begins, manipulation of colonies should be discontinued, the hives closed, and, if necessary, the entrances contracted as far as the weather will permit. If brush is thrown in front of the entrance, robbers are less likely to attempt entering. At all times honey which has been removed from the hives should be kept where no bees can get at it, so as not to incite robbing.

FEEDING.

During spring manipulations, in preparing bees for winter, and at other times it may be necessary to feed bees for stimulation or to provide stores. *Honey from an unknown source should never be used*, for fear of introducing disease, and sirup made of granulated sugar is cheapest and best for this purpose. The cheaper grades of sugar or molasses should never be used for winter stores. The proportion of sugar to water depends on the season and the purpose of the feeding. For stimulation a proportion of one-fourth to one-third sugar by volume is enough, and for fall feeding, especially if rather late, a solution containing as much sugar as it will hold when cold is best. There seems to be little advantage in boiling the sirup. Tartaric acid in small quantity may be added for the purpose of changing part of the cane sugar to invert sugar, thus retarding granulation. The medication of sirup as a preventive or cure of brood disease is often practiced, but it has not been shown that such a procedure is of any value. If honey is fed, it should be diluted somewhat, the amount of dilution depending on the season. If robbing is likely to occur, feeding should be done in the evening.

Numerous feeders are on the market, adapted for different purposes and methods of manipulation (figs. 17, 18, 19). A simple feeder can be made of a tin pan filled with excelsior or shavings (fig. 20). This is filled with sirup and placed on top of the frames in a super or hive body. It is advisable to lean pieces of wood on the pan as runways for the bees, and to attract them first to the sirup, either by mixing in a little honey or by spilling a little sirup over the frames and sticks.

It may be stated positively that it does not pay financially, or in any other way, to feed sugar sirup to be stored in sections and sold as comb honey. Of course, such things have been tried, but the consumption of sugar during the storing makes the cost greater than the value of pure floral honey.

SPRING MANAGEMENT.

The condition of a colony of bees in the early spring depends largely upon the care given the bees the preceding autumn and in the method of wintering. If the colony has wintered well and has a good prolific queen, preferably young, the chances are that it will become strong in time to store a good surplus when the honey flow comes.

The bees which come through the winter, reared the previous autumn, are old and incapable of much work. As the season opens they go out to collect the early nectar and pollen, and also care for the brood. The amount of brood is at first small, and as the new workers emerge they assist in the brood rearing so that the extent of the brood can be gradually increased until it reaches its maximum about the beginning of the summer. The old bees die off rapidly. If brood rearing does not continue late in the fall, so that the colony goes into winter with a large percentage of young bees, the old bees may die off in the spring faster than they are replaced by emerging brood. This is known as "spring dwindling." A preventive remedy for this may be applied by feeding, if necessary, the autumn before, or keeping up brood rearing as late as possible by some other means.

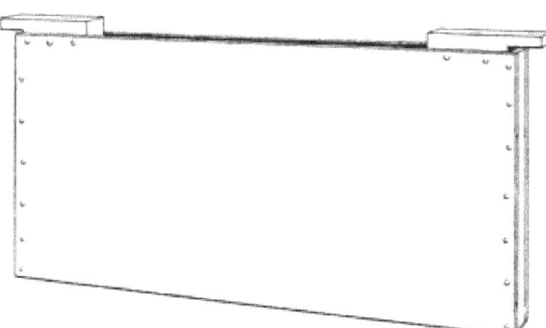

FIG. 17.—Division-board feeder to be hung in hive in place of frame.

If spring dwindling begins, however, it can be diminished somewhat by keeping the colony warm and by stimulative feeding, so that all the energy of the old bees may be put to the best advantage in rearing brood to replace those drying off. The size of the brood chamber can also be reduced to conserve heat.

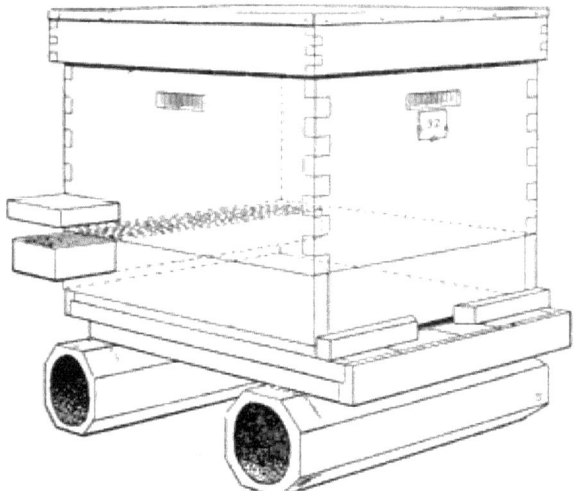

FIG. 18.—Feeder set in collar under hive body.

It sometimes happens that when a hive is examined in the spring the hive body and combs are spotted with brownish yellow excrement. This is an evidence of what is commonly called "dysentery." The cause of this trouble is long-continued confinement with a poor quality of honey for food. Honeydew honey and some of the inferior floral honeys contain a relatively large percentage of material which bees can not digest, and, if they are not able to fly for some time, the intestines become clogged with fæcal matter and a diseased condition results. Worker bees never normally deposit their fæces in the hive. The obvious preventive for this is to provide the colony with good honey or sugar sirup the previous fall. "Dysentery" frequently entirely destroys colonies, but if the bees can pull through until warm days permit a cleansing flight they recover promptly.

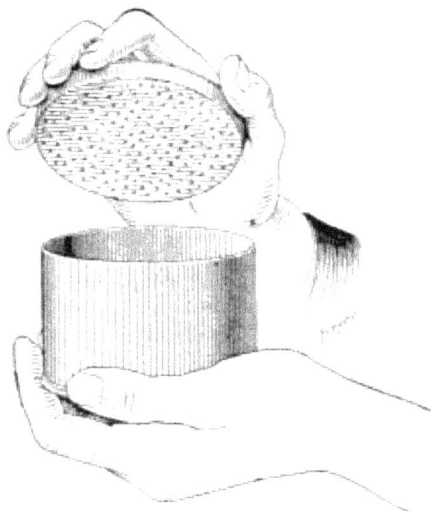

FIG. 19.—"Pepper-box" feeder for use on top of frames.

Bees should not be handled in the early spring any more than necessary, for to open a hive in cool weather wastes heat and may even kill the brood by chilling. The hive should be kept as warm as possible in early spring as an aid to brood rearing. It is a good practice to wrap hives in black tar paper in the spring, not only that it may aid in conserving the heat of the colony, but in holding the suns heat rays as a help to the warmth of the hive. This wrapping should be put on as soon as an early examination has shown the colony to be in good condition, and there need be no hurry in taking it off. A black wrapping during the winter is not desirable, as it might induce brood rearing too early and waste the strength of the bees.

As a further stimulus to brood rearing, stimulative feeding of sugar sirup in early spring may be practiced. This produces much the same effect as a light honey flow does and the results are often good. Others prefer to give the bees such a large supply of stores in the fall that when spring comes they will have an abundance for brood rearing, and it will not be necessary to disturb them in cool weather. Both ideas are good, but judicious stimulative feeding usually more than pays for the labor. Colonies should be fed late in the day, so that the bees will not fly as a result of it, and so that robbing will not be started. When the weather is warmer and more settled the brood cluster may be artificially enlarged by spreading the frames so as to insert an empty comb in the middle. The bees will attempt to cover all the brood that they already had, and the queen will at once begin laying in the newly inserted comb, thus making a great increase in the brood. This practice is desirable when carefully done, but may lead to serious results if too much new brood is produced. A beginner had better leave the quantity of brood to the bees.

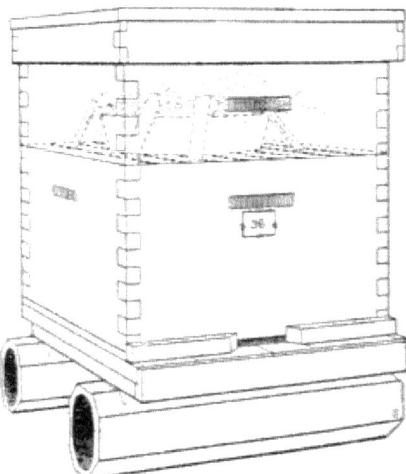

FIG. 20.—Pan in super arranged for feeding.

It is desirable early in the season, before any preparations are made for swarming, to go through the apiary and clip one wing of each queen (see p. 30). This should be done before the hive becomes too populous. It is perhaps best to clip queens as they are introduced, but some colonies may rear new ones without the knowledge of the owner, and a spring examination will insure no escaping swarms. The beginner should perhaps be warned not to clip the wings of a virgin queen.

Queens sometimes die during the winter and early spring, and since there is no brood from which the bees can replace them, the queenless colonies are "hopelessly queenless." Such colonies are usually restless and are not active in pollen gathering. If, on opening a colony, it is found to be without a queen and reduced in numbers, it should be united with another colony by smoking both vigorously and caging the queen in the queen-right colony for a day or two to prevent her being killed. A frame or two of brood may be added to a queenless colony, not only to increase its strength, but to provide young brood from which they can rear a queen Bee keepers in the North can frequently buy queens from southern breeders early in the spring and naturally this is better than leaving the colony without a queen until the bees can rear one, as it is important that there be no stoppage in brood rearing at this season.

SWARM MANAGEMENT AND INCREASE.

The excessive rearing of brood at the wrong season or increase in the number of colonies greatly reduces the surplus honey crop by consumption. The ideal to which all progressive bee keepers work, when operating simply for honey, is to stimulate brood rearing to prepare bees for gathering, to retard breeding when it is less desirable, and to prevent swarming. Formerly the measure of success in beekeeping was the amount of increase by swarming, but this is now recognized as being quite the contrary of success.

The stimulation of brood rearing in the spring, however, makes it more likely that swarming will occur; so that the operator must counteract the tendency to swarm. This is especially true in comb honey production. Very few succeed in entirely preventing swarming, but by various methods the situation can be largely controlled.

When a swarm issues, it usually first settles on a limb of a tree or bush near the apiary. It was formerly common to make a noise by beating pans or ringing bells in the belief that this causes the swarm to settle. There is no foundation for such action on the part of the bee keeper. If the bees alight on a small limb that can be spared it may simply be sawed off and the bees carried to the hive and thrown on a sheet or hive cover in front of the entrance. If the limb can not be cut, the swarm can be shaken off into a box or basket on a pole and hived. If the bees light on the trunk of a tree or in some inaccessible place they can first be attracted away by a comb, preferably containing unsealed brood. In these manipulations it is not necessary to get all the bees, but if the queen is not with those which are put into the hive the bees will go into the air again and join the cluster.

If a queen is clipped as recommended under "Spring management" (p. 29) the swarm will issue just the same, but the queen, not being able to fly, will simply wander about on the ground in front of the hive, where she can be caught and caged. The parent colony can then be removed to a new stand and a new hive put in its place. The bees will soon return and the queen can be freed among them as they enter. The field bees on returning will enter the new hive with the swarm, thus decreasing still more the parent colony and making a second swarm less probable. To make sure of this, however, all queen cells except one good one can be removed soon after the swarm issues. Another method of preventing second swarms is to set the old hive beside the swarm and in a week move the old hive to another place. The field bees of the parent colony then join the swarm and the parent colony is so much reduced that a second swarm does not issue.

To hold a swarm it is desirable to put one frame containing healthy unsealed brood in the new hive. The other frames may contain full sheets or starters of foundation. Usually comb honey supers or surplus bodies for extracting frames will have been put on before swarming occurs. These are given to the swarm on the old stand and separated from the brood chamber by queen-excluding perforated zinc. In three or four days the perforated zinc may be removed if desired.

When clipping the queen's wing is not practiced, swarms may be prevented from leaving by the use of queen traps of perforated zinc (fig. 6). These allow the workers to pass out, but not drones or queens, which, on leaving the entrance, pass up to an upper compartment from which they can not return. These are also used for keeping undesirable drones from escaping, and the drones die of starvation. When a swarm issues from a hive provided with a queen trap, the queen goes to the upper compartment and remains there until released by the bee keeper. The workers soon return to the hive. When the operator discovers the queen outside, the colony may be artificially swarmed to prevent another attempt at natural swarming. A queen trap should not be kept on the hive all the time for fear the old queen may be superseded and the young queen prevented from flying out to mate.

ARTIFICIAL SWARMING.

If increase is desired, it is better to practice some method of artificial swarming and to forestall natural swarming rather than be compelled to await the whims of the colonies. The situation should be under the control of the bee keeper as much as possible. The bees, combs, and brood may be divided into two nearly equal parts and a queen provided for the queenless portion; or small colonies, called nuclei, may be made from the parent colony, so reducing its strength that swarming is not attempted. These plans are not as satisfactory as shaken swarms, since divided colonies lack the vigor of swarms.

A good method of artificially swarming a colony is to shake most of the bees from the combs into another hive on the old stand with starters (narrow strips) of foundation. The hive containing the brood with some bees still adhering is then moved to a new location. If receptacles for surplus honey have been put on previously, as they generally should be, they should now be put over the artificial swarm separated from the brood compartment by perforated zinc.

This method of artificially swarming (usually called by bee keepers "shook" swarming) should not be practiced too early, since natural swarming may take place later. The colony should first have begun its preparations for swarming. The method is particularly useful in comb honey production. The bees may be prevented from leaving the hive by the use of a drone trap (fig. 6) or by putting in one frame containing unsealed brood. Some bee keepers prefer using full sheets of foundation or even drawn combs for the artificial swarm, but narrow strips of foundation have some advantages. By using narrow strips the queen has no cells in which to lay eggs for a time, thus reducing brood rearing, but, since by the time artificial swarming is practiced the profitable brood rearing is usually over, this is no loss but rather a gain. There are also in the brood compartment no cells in which the gathering workers can deposit fresh honey, and they consequently put it in the supers. Gradually the combs below are built out and brood rearing is increased. Later the colony is allowed to put honey in the brood combs for its winter supply. If no increase is desired, the bees which emerge from the removed brood combs may later be united with the artificial swarm and by that time there will usually be little danger of natural swarming.

Artificial swarming can readily be combined with the shaking treatment for bee diseases, thus accomplishing two objects with one manipulation. If disease is present in the parent colony, only strips of foundation should be used and the colony should be confined to the hive until a queen and drone trap and not with a frame of brood.

PREVENTION OF SWARMING.

Unless increase is particularly desired, both natural and artificial swarming should be done away with as far as possible, so that the energy of the bees shall go into the gathering of honey. Since crowded and overheated hives are particularly conducive to swarming, this tendency may be largely overcome by giving plenty of ventilation and additional room in the hive. Shade is also a good preventive of swarming. Extra space in the hive may be furnished by adding more hive bodies and frames or by frequent extracting, so that there may be plenty of room for brood rearing and storage at all times. These manipulations are, of course, particularly applicable to extracted honey production.

To curb the swarming impulse frequent examinations of the colonies (about every week or 10 days during the swarming season) for the purpose of cutting out queen cells is a help, but this requires considerable work, and since some cells may be overlooked, and particularly since it frequently fails in spite of the greatest care, it is not usually practiced. Requeening with young queens early in the season, when possible, generally prevents swarming.

Swarming is largely due to crowded brood chambers, and since eggs laid immediately before and during the honey flow do not produce gatherers, several methods have been tried of reducing the brood. The queen may either be entirely removed or be caged in the hive to prevent her from laying. In either event the bees will usually build queen cells to replace her, and these must be kept cut out. These plans would answer the purpose very well were it not for the fact that queenless colonies often do not work vigorously. Under most circumstances these methods can not be recommended. A better method is to remove brood about swarming time and thus reduce the amount. There are generally colonies in the apiary to which frames of brood can be given to advantage.

In addition to these methods various nonswarming devices have been invented, and later a nonswarming hive so constructed that there is no opportunity for the bees to form a dense cluster. The breeding of bees by selecting colonies with less tendency to swarm has been suggested.

On the whole, the best methods are the giving of plenty of room, shade, and ventilation to colonies run for extracted honey; and ventilation, shade, and artificial swarming of colonies run for comb honey. Frequent requeening (about once in two years) is desirable for other reasons, and requeening before swarming time helps in the solution of that difficulty,

PREPARATION FOR THE HARVEST.

An essential in honey production is to have the hive overflowing with bees at the beginning of the honey flow, so that the field force will be large enough to gather more honey than the bees need for their own use. To accomplish this, the bee keeper must see to it that brood rearing is heavy some time before the harvest, and he must know accurately when the honey flows come, so that he may time his manipulations properly. Brood rearing during the honey flow usually produces bees which consume stores, while brood reared before the flow furnishes the surplus gatherers. The best methods of procedure may be illustrated by giving as an example the conditions in the white clover region.

In the spring the bees gather pollen and nectar from various early flowers, and often a considerable quantity from fruit bloom and dandelions. During this time brood rearing is stimulated by the new honey, but afterwards there is usually a period of drought when brood rearing is normally diminished or not still more increased as it should be. This condition continues until the white clover flow comes on, usually with a rush, when brood rearing is again augmented. If such a condition exists, the bee keeper should keep brood rearing at a maximum by stimulative feeding during the drought. When white clover comes in bloom he may even find it desirable to prevent brood rearing to turn the attention of his bees to gathering.

A worker bee emerges from its cell 21 days after the egg is laid, and it usually begins field work in from 14 to 17 days later. It is evident, therefore, that an egg must be laid five weeks before the honey flow to produce a gatherer. Since the flow continues for some time and since bees often go to the field earlier than 14 days, egg laying should be pushed up to within two or three weeks of the opening of the honey flow. In addition to stimulative feeding, the care of the colony described under the heading of "Spring management" (p. 26) will increase brood production.

THE PRODUCTION OF HONEY.

The obtaining of honey from bees is generally the primary object of their culture. Bees gather nectar to make into honey for their own use as food, but generally store more than they need, and this surplus the bee keeper takes away. By managing colonies early in the spring as previously described the surplus may be considerably increased. The secret of maximum crops is to "Keep all colonies strong."

Honey is gathered in the form of nectar secreted by various flowers, is transformed by the bees, and stored in the comb. Bees also often gather a sweet liquid called "honeydew," produced by various scale insects and plant-lice, but the honeydew honey made from it is quite unlike floral honey in flavor and composition and should not be sold for honey. It is usually unpalatable and should never be used as winter food for bees, since it usually causes dysentery (p. 40). When nectar or honeydew has been thickened by evaporation and otherwise changed, the honey is sealed in the cells with cappings of beeswax.

It is not profitable to cultivate any plant solely for the nectar which it will produce, but various" plants, such as clovers, alfalfa, and buckwheat, are valuable for other purposes and are at the same time excellent honey plants; their cultivation is therefore a benefit to the bee keeper. It is often profitable to sow some plant on waste land; sweet clovers are often used in this way. The majority of honey-producing plants are wild, and the bee keeper must largely accept the locality as he finds it and manage his apiary so as to get the largest possible amount of the available nectar. Since bees often fly as far as 2 or 3 miles to obtain nectar, it is obvious that the bee keeper can rarely influence the nectar supply appreciably. Before deciding what kind of honey to produce the bee keeper should have a clear knowledge of the honey resources of his locality and of the demands of the market in which he will sell his crop. If the bulk of the honey is dark, or if the main honey flows are slow and protracted, it will not pay to produce comb honey, since the production of fancy comb honey depends on a rapid flow. The best localities for comb honey production are in the northern part of the United States east of the Mississippi River, where white clover is a rapid and abundant yielder. Other parts of the United States where similar conditions of rapidity of flow exist are also good. Unless these favorable conditions are present it is better to produce extracted Honey.

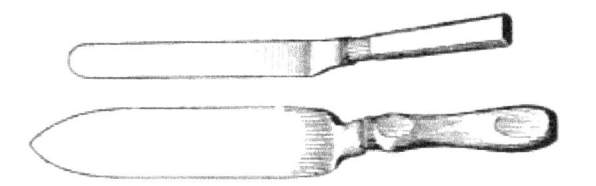

FIG. 21.—Knives for uncapping honey.

EXTRACTED HONEY.[1]

[1] For farther discussion of the production and care of extracted honey, see Bulletin 75, Part I, Bureau of Entomology.

Extracted honey is honey which has been removed by means of centrifugal force from the combs in which the bees stored it. While it is possible to adulterate extracted honey by the addition of cheap sirups, this is rarely done, perhaps largely on account of the possibility of detection. It may be said to the credit of bee keepers as a class that they have always opposed adulteration of honey.

In providing combs for the storage of honey to be extracted the usual practice is to add to the top of the brood chamber one or more hive bodies just like the one in which brood is reared, and fill these with frames. If preferred, shallower frames with bodies of proper size may be used, but most honey extractors are made for full-size frames. The surplus bodies should be put on in plenty of time to prevent the crowding of the brood chamber, and also to act as a preventive of swarming.

Honey for extracting should not be removed until it is well ripened and a large percentage of it capped. It is best, however, to remove the crop from each honey flow before another heavy producing plant comes into bloom, so that the different grades of honey may be kept separate. It is better to extract while honey is still coming in, so that the bees will not be apt to rob. The extracting should be done in a building, preferably one provided with wire-cloth at the windows (p. 9).

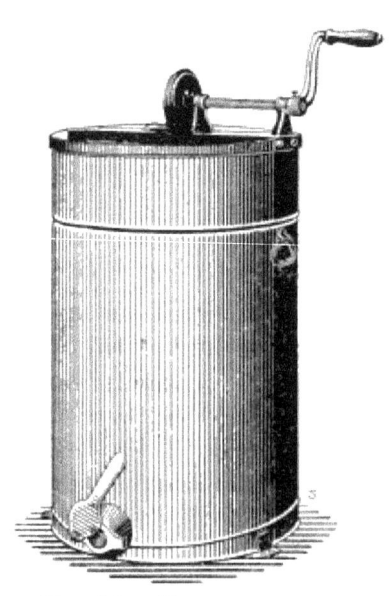

FIG. 22.—Honey extractor.

The frames containing honey to be extracted are removed from the hive, the cappings cut off with a sharp, warm knife (fig. 21) made specially for this purpose, and the frames are then put into the baskets of the honey extractor (fig. 22) . By revolving these rapidly the honey is thrown out of one side. The basket is then reversed and the honey from the other side is removed. The combs can then be returned to the bees to be refilled, or if the honey flow is over, they can be returned to the bees to be cleaned and then removed and stored until needed again. This method is much to be preferred to mashing the comb and straining out the honey, as was formerly done.

In large apiaries special boxes to receive cappings, capping melters to render the cappings directly into wax, and power-driven extractors are often used. These will be found listed in supply catalogues.

The extracted honey is then strained and run into vessels. It is advisable not to put it in bottles at once, but to let it settle in open vessels for a time, so that it can be skimmed. Most honeys will granulate and become quite hard if exposed to changes of temperature, and to liquefy granulated extracted honey it should be heated in a water bath. Never heat honey directly over a stove or flame, as the flavor is thereby injured. The honey should never be heated higher than 160° F. unless it is necessary to sterilize it because of contamination by disease.

Extracted honey is put up in bottles or small tin cans for the retail trade, and in 5-gallon square tin cans or barrels for the wholesale market. Great care must be exercised if barrels are used, as honey will absorb moisture from the wood, if any is present, and cause leakage. The tin package is much to be preferred in most cases. In bottling honey for retail trade, it will well repay the bee keeper or bottler to go to considerable expense and trouble to make an attractive package, as the increased price received will more than compensate for the increased labor and expense. Honey should be heated to 160° F. and kept there for a time before bottling, and the bottle should be filled as full as possible and sealed hermetically.

Granulated honey.—Some honeys, such as alfalfa, granulate quickly after being extracted. Such honeys are sometimes allowed to granulate in large cans and the semisolid mass is then cut into 1-pound bricks like a butter print and wrapped in paraffin paper. It may be put into paraffined receptacles before granulation, if desired. There is always a ready market for granulated honey, since many people prefer it to the liquid honey.

COMB HONEY.

Comb honey is honey as stored in the comb by the bees, the size and shape being determined by the small wooden sections provided by the bee keeper. Instead of having comb in large frames in which to store surplus honey, the bees are compelled to build comb in the sections and to store honey there (fig. 2). A full section weighs about 1 pound; larger ones are rarely used. By the use of modern sections and foundation the comb honey now produced is a truly beautiful, very uniform product, so uniform in fact that it is often charged that it must be artificially manufactured. The purchaser of a section of comb honey may be absolutely certain, however, that he is obtaining a product of the bees, for never has anyone been able to imitate the bees' work successfully. To show their confidence in the purity of comb honey, the National Bee Keepers' Association offers $1,000 for a single pound of artificial comb filled with an artificially prepared sirup, which is at all difficult of detection.

There are several different styles of sections now in use, the usual sizes being 4¼ inches square and 4 by 5 inches. There are also two methods of spacing, so that there will be room for the passage of bees from the brood chamber into the sections and from one super of sections to another. This is done either by cutting "bee ways" in the sections and using plain flat separators or by using "no bee-way" or plain sections and using "fences"—separators with cleats fastened on each side, to provide the bee space. To describe all the different "supers" or bodies for holding sections would be impossible in a bulletin of this size, and the reader must be referred to catalogues of dealers in beekeeping supplies. Instead of using regular comb honey supers, some bee keepers use wide frames to hold two tiers of sections. It is better, however, to have the supers smaller, so that the bees may be crowded more to produce full sections. To overcome this difficulty, shallow wide frames holding one tier of sections may be used. The majority of bee keepers find it advisable to use special comb honey supers.

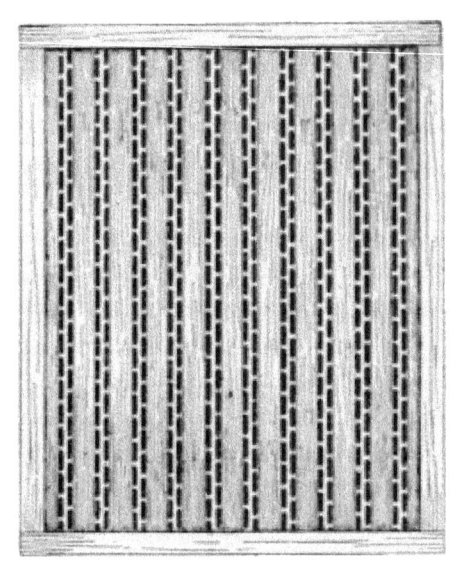

FIG. 23.—Perforated zinc queen excluder.

In producing comb honey it is even more necessary to know the plants which produce surplus honey, and just when they come in bloom, than it is in extracted honey production. The colony should be so manipulated that the maximum field force is ready for the beginning of the flow. This requires care in spring management, and, above all, the prevention of swarming. Supers should be put on just before the heavy flow begins. A good indication of the need of supers is the whitening of the brood combs at the top. If the bees are in two hive bodies they should generally be reduced to one, and the frames should be filled with brood and honey so that as the new crop comes in the bees will carry it immediately to the sections above. If large hives are used for the brood chamber it is often advisable to remove some of the frames and use a division board to crowd the bees above. To prevent the queen from going into the sections to lay, a sheet of perforated zinc (fig. 23) may be put between the brood chamber and the super (fig. 2).

It is often difficult to get bees to begin work in the small sections, but this should be brought about as soon as possible to prevent loss of honey.

If there are at hand some sections which have been partly drawn the previous year, these may be put in the super with the new sections as "bait." Another good plan is to put a shallow extracting frame on either side of the sections. If a few colonies in the apiary that are strong enough to go above refuse to do so, lift supers from some colonies that have started to work above and give them to the slow colonies. The super should generally be shaded somewhat to keep it from getting too hot. Artificial swarming will quickly force bees into the supers.

To produce the finest quality of comb honey full sheets of foundation should be used in the sections. Some bee keepers use nearly a full sheet hung from the top of the section and a narrow bottom starter. The use of foundation of worker-cell size is much preferred.

When one super becomes half full or more and there are indications that there will be honey enough to fill others, the first one should be raised and an empty one put on the hive under it. This tiering up can be continued as long as necessary, but it is advisable to remove filled sections as soon as possible after they are nicely capped, for they soon become discolored and less attractive. Honey removed immediately after capping finds a better market, but if left on the hive even until the end of the summer the quality of the honey is improved. A careful watch must be kept on the honey flow, so as to give the bees only enough sections to store the crop. If this is not done a lot of unfinished sections will be left at the end of the flow. Honeys from different sources should not be mixed in the sections, as it usually gives the comb a bad appearance

To remove bees from sections, the super may be put over a bee escape so that the bees can pass down but can not return, or the supers may be removed and covered with a wire-cloth-cone bee escape.

FIG. 24.—Shipping case for comb honey.

After sections are removed the wood should be scraped free of propolis (bee glue) and then packed in shipping cases (fig. 24) for the market. Shipping cases to hold 12, 24, or 48 sections, in which the various styles of sections fit exactly, are manufactured by dealers in supplies. In shipping these cases, several of them should be put in a box or crate packed in straw and paper and handles provided to reduce the chances of breakage. When loaded in a freight car the combs should be parallel with the length of the car.

In preparing comb honey for market it should be carefully graded so that the sections in each shipping case are as uniform as possible. Nothing will more likely cause wholesale purchasers to cut the price than to find the first row of sections in a case fancy and those behind of inferior grade. Grading rules have been adopted by various bee keepers' associations or drawn up by honey dealers. The following sets of rules are in general use:

EASTERN GRADING RULES FOR COMB HONEY.

Fancy.—All sections -well filled; combs straight; firmly attached to all four sides; the combs unsoiled by travel, stain, or otherwise; all the cells sealed except an occasional one; the outside surface of the wood well scraped of propolis.

A No. 1.—All sections well filled except the row of cells next to the wood; combs straight; one-eighth part of comb surface soiled, or the entire surface slightly soiled; the outside surface of the wood well scraped of propolis.

No. 1.—All sections well filled except the row of cells next to the wood; combs comparatively even; one-eighth part of comb surface soiled, or the entire surface slightly soiled.

No. 2.—Three-fourths of the total surface must be filled and sealed.

No. 3.—Must weigh at least half as much as a full-weight section.

In addition to this the honey is to be classified according to color, using the terms white, amber, and dark; that is, there will be "Fancy White," "No. 1 Dark," etc.

NEW COMB HONEY GRADING RULES ADOPTED BY THE COLORADO STATE BEE KEEPERS' ASSOCIATION.

No. 1 White.—Sections to be well filled and evenly capped, except the outside row, next to the wood ; honey white or slightly amber, comb and cappings white, and not projecting beyond the wood; wood to be well cleaned; cases of separatored honey to average 21 pounds net per case of

24 sections; no section in this grade to weigh less than 13½ ounces. Cases of half-separatored honey to average not less than 22 pounds net per case of 24 sections. Cases of unseparatored honey to average not less than 23 pounds net per case of 24 sections.

No. 1 Light Amber.—Sections to be well filled and evenly capped, except the outside row next to the wood; honey white or light amber; comb and cappings from white to off color, but not dark; comb not projecting beyond the wood; wood to be well cleaned. Cases of separatored honey to average 21 pounds net per case of 24 sections; no section in this grade to weigh less than 13½ ounces. Cases of half-separatored honey to average not less than 22 pounds net per case of 24 sections. Cases of unseparatored honey to average not less than 23 pounds net per case of 24 sections.

No. 2.—This includes all white honey, and amber honey not included in the above grades; sections to be fairly well filled and capped, no more than 25 uncapped cells, exclusive of outside row, permitted in this grade; wood to be well cleaned; no section in this grade to weigh less than 12 ounces. Cases of separatored honey to average not less than 19 pounds net. Cases of half-separatored honey to average not less than 20 pounds net per case of 24 sections. Cases of unseparatored honey to average not less than 21 pounds net per case of 24 sections.

THE PRODUCTION OF WAX.

Beeswax, which is secreted by the bees and used by them for building their combs, is an important commercial product. There are times in almost every apiary when there are combs to be melted up, and it pays to take care of even scraps of comb and the cappings taken off in extracting. A common method of taking out the wax is to melt the combs in a solar wax extractor. This is perhaps the most feasible method where little wax is produced, but considerable wax still remains in old brood combs after such heating. Various wax presses are on the market, or one can be made at home. If much wax is produced, the bee keeper should make a careful study of the methods of wax extraction, as there is usually much wax wasted even after pressing.

PREPARATIONS FOR WINTERING.

After the main honey flow is over the management must depend on what may be expected later in the season from minor honey flows. If no crop is to be expected, the colony may well be kept only moderately strong, so that there will not be so many consumers in the hive.

In localities where winters are severe and breeding is suspended for several months great care should be taken that brood rearing is rather active during the late summer, so that the colony may go into winter with plenty of young bees. In case any queens show lack of vitality they should be replaced early, so that the bees will not become queenless during the winter.

The important considerations in wintering are plenty of young bees, a good queen, plenty of stores of good quality, sound hives, and proper protection from cold and dampness.

If, as cold weather approaches, the bees do not have stores enough, they must be fed. Every colony should have from 25 to 40 pounds, depending on the length of winter and the methods of wintering. It is better to have too much honey than not enough, for what is left is good next season. If feeding is practiced, honey may be used, but sirup made of granulated sugar is just as good and is perfectly safe. If honey is purchased for feeding, great care should be taken that it comes from a healthy apiary, otherwise the apiary may be ruined by disease. *Never feed honey bought on the open market.* The bees should be provided with stores early enough so that it will not be necessary to feed or to open the colonies after cold weather comes on. Honeydew honey should not be left in the hives, as it produces "dysentery." Some honeys are also not ideal for winter stores. Those which show a high percentage of gums (most tree honeys) are not so desirable, but will usually cause no trouble.

In wintering out of doors the amount of protection depends on the severity of the winter. In the South no packing is necessary, and even in very cold climates good colonies with plenty of stores can often pass the winter with little protection, but packing and protection make it necessary for the bees to generate less heat, and consequently they consume less stores and their vitality is not reduced. Dampness is probably harder for bees to withstand than cold, and when it is considered that bees give off considerable moisture, precautions should be taken that as it condenses it does not get on the cluster. An opening at the top would allow the moisture to pass out, but it would also waste heat, so it is better to put a mat of burlap or other absorbent material on top of the frames. The hive may also be packed in chaff, leaves, or other similar dry material to diminish the loss of heat. Some hives are made with double walls, the space being filled with chaff; these are good for outdoor wintering. The hive entrance should be lower than any other part of the hive, so that any condensed moisture may run out. The hives should be sound and the covers tight and waterproof.

Entrances should be contracted in cold weather not only to keep out cold wind, but to prevent mice from entering. There should always be enough room, however, for bees to pass in and out if warmer weather permits a flight.

In the hands of experienced bee keepers cellar wintering is very successful, but this method requires careful study. The cellar must be dry and so protected that the temperature never varies more than from 40 to 45° F.; 43° F. seems to be the optimum temperature. The ventilation must be good or the bees become fretful. Light should not be admitted to the cellar, and consequently some means of indirect ventilation is necessary.

Cellar wintering requires the consumption of less honey to maintain the proper temperature in the cluster and is therefore economical. Bees so wintered do not have an opportunity for a cleansing flight, often for several months, but the low consumption makes this less necessary. Some bee keepers advocate carrying the colonies out a few times on warm days, but it is not fully established whether this is entirely beneficial and it is usually not practiced.

The time for putting colonies in the cellar is a point of dispute, and practice in this regard varies considerably. They should certainly be put in before the weather becomes severe and as soon as they have ceased brood rearing. The time chosen may be at night when they are all in the hive, or on some chilly day.

The hives may be piled one on top of the other, the lower tier raised a little from the floor. The entrances should not be contracted unless the colony is comparatively weak. It is usually not considered good policy to close the entrances with ordinary wire cloth, as the dead bees which accumulate more or less on the bottom boards may cut off ventilation, and the entrance should be free so that these may be cleaned out.

It is, however, good policy to cover the entrance with wire-cloth having three meshes to the inch to keep out mice.

The time of removing bees from the cellar is less easily determined than that of putting them in. The colonies may be removed early and wrapped in *black* tar paper or left until the weather is settled. If the weather is very warm and the bees become fretful, the cellar must either be cooled or the bees removed. Some bee keepers prefer to remove bees at night, so that they can recover from the excitement and fly from the hive normally in the morning. One of the chief difficulties is to prevent the bees from getting into the wrong hives after their first flights. They often "drift" badly with the wind, and sometimes an outside row will become abnormally strong, leaving other colonies weak.

The night before the bees are removed from the cellar it is good practice to leave the cellar doors and windows wide open.

DISEASES AND ENEMIES.

There are two infectious diseases of the brood of bees which cause great losses to the beekeeping industry of the United States. These are known as American foul brood and European foul brood. Both of these diseases destroy colonies by killing the brood, so that there are not enough young bees emerging to take the place of the old adult bees as these die from natural causes. The adult bees are not attacked by either disease. In the hands of careful bee keepers both diseases may be controlled, and this requires careful study and constant watching. In view of the fact that these diseases are now widely distributed throughout the United States, every bee keeper should read the available literature on the subject, so that if disease enters his apiary he may be able to recognize it before it gets a start. The symptoms and the treatment recommended by this department are given in another publication which will be sent free on request.[2]

[2] Farmers' Bulletin No. 442. "The Treatment of Bee Diseases."

It is difficult for a bee keeper to keep his apiary free from disease if others about him have diseased colonies which are not properly treated. The only way to keep disease under control is for the bee keepers in the neighborhood to cooperate in doing everything possible to stamp out disease as soon as it appears in a single colony. The progressive bee keeper who learns of disease in his neighborhood should see to it that the other bee keepers around him are supplied with literature describing symptoms and treatment, and should also try to induce them to unite in eradicating the malady. Since it is so often impossible to get all of the bee keepers in a community to treat infected colonies properly and promptly, it is desirable that the States pass laws providing for the inspection of apiaries and granting to the inspector the power to compel negligent bee keepers to treat diseased colonies so that the property of others may not be endangered and destroyed. This has been done in a number of States, but there are still some where the need is great and in which no such provision has been made. When no inspection is provided, bee keepers should unite in asking for such protection, so that the danger to the industry may be lessened.

In case there is an inspector for the State or county, he should be notified as soon as disease is suspected in the neighborhood. Some bee keepers hesitate to report disease through fear that the inspector will destroy their bees or because they feel that it is a disgrace to have disease in the apiary. There is no disgrace in having colonies become diseased; the discredit is in not treating them promptly. The inspectors are usually, if not universally, good practical bee keepers who from a wide experience are able to tell what should be done in individual cases to give the best results with the least cost in material and labor. They do not destroy colonies needlessly, and, in fact, they all advocate and teach treatment.

The brood diseases are frequently introduced into a locality by the shipping in of diseased colonies; or, more often, the bees get honey from infected colonies which is fed to them, or which they rob, from discarded honey cans. It is decidedly dangerous to purchase honey on the market, with no knowledge of its source, to be used in feeding bees. Many outbreaks of disease can be traced to this practice (see "Feeding," p. 26). It is difficult to prevent bees from getting contaminated honey accidentally. If colonies are purchased, great care should be taken that there is no disease present. Whenever possible, colonies should be purchased near at home, unless disease is already present in the neighborhood.

There are other diseased conditions of the brood, known to bee keepers as "pickle brood," but these can usually be distinguished from the two diseases previously mentioned. The so-called "pickle brood" is not contagious and no treatment is necessary. Bees also suffer from "dysentery," which is discussed in the earlier part of this bulletin, and from the so-called "paralysis," a disease of adult bees. No treatment for the latter disease can as yet be recommended as reliable. The sprinkling of powdered sulphur on the top bars of frames or at the entrance is sometimes claimed to be effective, but under what circumstances it is beneficial is unknown.

A number of insects, birds, and mammals must be classed as enemies of bees, but of these the two wax moths, and ants, are the only ones of importance. There are two species of moth, the larger wax moth (*Galleria mellonella* L.), and the lesser wax moth (*Achroia grisella* Fab.), the larvæ of which destroy combs by burrowing through them.[3] Reports are frequently received in the department that the larvæ of these moths (usually the larger species) are destroying colonies of bees. It may be stated positively that moths do not destroy strong, healthy colonies in good hives, and if it is supposed that they are causing damage the bee keeper should carefully study his colonies to see what other trouble has weakened them enough for the moths to enter. Queenlessness, lack of stores, or some such trouble may be the condition favorable to the entrance of the pest, but a careful examination should be made of the brood to see whether there is any evidence of disease. This is the most frequent cause of the cases of moth depredation reported to this department. Black bees are less capable of driving moth larvæ out, but, even with these bees, strong colonies rarely allow them to remain. The observance of the golden rule of beekeeping, "Keep all colonies strong," will solve the moth question unless disease appears.

[3] Bee keepers refer to these insects as "moths," "wax moths," "bee moths," "millers," "wax worms," "honey moths," "moth worms," "moth millers," and "grubs." The last six terms are not correct.

Moth larvæ often destroy combs stored outside the hive. To prevent this the combs may be fumigated with sulphur fumes or bisulphid of carbon in tiers of hives or in tight rooms. If bisulphid of carbon is used, great care should be taken not to bring it near a flame, as it is highly inflammable. Combs should be stored in a dry, well-ventilated, light room.

In the warmer parts of the country ants are often a serious pest. They may enter the hive for protection against changes of temperature, or to prey on the honey stores or the brood. The usual method of keeping them out is to put the hive on a stand, the legs of which rest in vessels containing water or creosote. Another method is to wrap a tape soaked in corrosive sublimate around the bottom board.

GENERAL INFORMATION.

For the purpose of answering numerous questions which are asked of this department the following brief topics are included.

BREEDERS OF QUEENS.

There are a large number of bee keepers who make a business of rearing queens of good stock for sale. The queens are usually sent by mail. If poor stock is all that can be obtained locally, it is recommended that such colonies be purchased and the queens removed and replaced with those obtained from a good breeder. This department can supply names of breeders, nearest the applicant, of any race raised in this country.

INTRODUCING QUEENS.

When queens are shipped by mail they usually come in cages (fig. 25) which can be used for introducing. If the colony to receive the new queen has one, she must be removed and the cage inserted between the frames. The small hole leading into the candy compartment is uncovered, and the bees gradually eat through and release the queen. If queens are reared at home, a similar cage may be used for introducing. In view of the fact that disease may be transmitted in mailing cages, it is always a wise precaution to remove the new queen and destroy the accompanying workers and the cage and its contents. The queen may then be put into a clean cage without worker bees, with candy known to be free from contamination (made from honey from healthy hives), and introduced in the regular way.

Queens sold by breeders are always mated unless otherwise specified, and consequently the colony in which they are introduced has no effect on her offspring. During the active season the bees in the colony are all the offspring of the new queen in about nine weeks. Three weeks is required for the previous brood to emerge (if the colony has not been queenless). and in six weeks after all the old brood emerges most of the workers from it will have died. Queens are usually sold according to the following classification:

"*Untested queen*"—one that has mated, but the race of the drone is not known.

"*Tested queen*"—one that has mated and has been kept only long enough to show, from the markings of her progeny, that she mated with a drone of her own race.

"*Breeding queen*"—a tested queen which has shown points of superiority, making her desirable for breeding purposes.

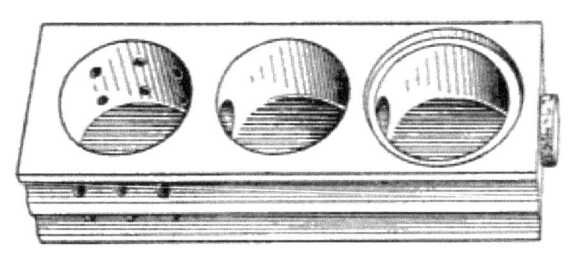

FIG. 25.—Queen mailing cage.

DEALERS IN BEE KEEPERS' SUPPLIES.

There are several manufacturers of supplies in this country who can furnish almost anything desired by the bee keeper. Some of them have agents in various parts of the country from whom supplies may be purchased, thus saving considerable in freight.

BEE KEEPERS' ASSOCIATIONS.

There are a large number of associations of bee keepers in all parts of the country, formed for the betterment of the industry, and a few associations which are organized to aid the members in purchasing supplies and in selling the crops. Of these the National Bee Keepers" Association is the largest. It helps its members in obtaining their legal rights, and aids in securing legislation for the furtherance of the industry. The annual conventions are held in different parts of the country, and copies of the proceedings are sent to the members. There are also numerous State, county, and town associations, some of which publish proceedings. The names of officers of the nearest associations or of the National Bee Keepers' Association will be sent from this department on request.

LAWS AFFECTING BEEKEEPING.

Disease inspection.—Various States have passed laws providing for the State or county inspection of apiaries for bee-disease control, and every bee keeper should get in touch with an inspector when disease is suspected, if one is provided. The inspectors are practical bee keepers who fully understand how to control the diseases, and are of great help in giving directions in this matter. The name of the inspector of any locality can usually be furnished, and this department is glad to aid bee keepers in reaching the proper officers.

Laws against spraying fruit trees while in bloom.—The spraying of fruit trees while in bloom is not now advised by economic entomologists, and to prevent the practice some States have passed laws making it a misdemeanor. Such spraying not only kills off honey bees, causing a loss to the bee keeper, but interferes with the proper pollination of the blossoms and is thus a detriment to the fruit grower. Bee keepers should do everything in their power to prevent the practice.

Laws against the adulteration of honey.—The national food and drugs act of 1906, and various State pure food laws, are a great aid to the bee keeper in preventing the sale of adulterated extracted honey as pure honey. Bee keepers can often aid in this work by reporting to the proper officials infringements of these laws which come to their notice.

When bees are a nuisance.—Some cities have passed ordinances prohibiting the keeping of bees in certain areas, but so far none has been able to enforce them. If bees are a nuisance in individual cases, the owner may be compelled to remove them. The National Bee Keepers' Association "will help any of its members in such cases, if they are in the right, as well as in cases where bees sting horses. Bee keepers should be careful not to locate bees where they can cause any trouble of this kind.

SUPPOSED INJURY OF CROPS BY BEES.

Bee keepers are often compelled to combat the idea that bees cause damage to fruit or other crops by sucking the nectar from the flower. This is not only untrue, but in many cases the bees are a great aid in the pollination of the flowers, making a good crop possible. A more frequent complaint is that bees puncture fruit and suck the juices. Bees never puncture sound fruit, but if the skin is broken by some other means bees will often suck the fruit dry. In doing it, however, they are sucking fruit which is already damaged. These and similar charges against the honey bee are prompted by a lack of information concerning their activities. Bees may, of course, become a nuisance to others through their stinging propensities, but bee keepers should not be criticized for things which their bees do not do.

JOURNALS AND BOOKS ON BEEKEEPING.

The progressive bee keeper will find it to his profit to subscribe for at least one journal devoted to beekeeping. Several of these are published in the United States. The names and addresses of such journals may usually be obtained from a subscription agent for periodicals, or from a supply dealer.

It will also be advantageous to read and study books on beekeeping, of which several are published in this country. These are advertised in journals devoted to beekeeping, or may usually be obtained through the local book dealer or through dealers in bee keepers' supplies.

PUBLICATIONS OF THE DEPARTMENT OF AGRICULTURE ON BEE KEEPING.[4]

[4] List revised to April 1, 1911. (VII.)

There are several publications of this department which are of interest to bee keepers, and new ones are added from time to time in regard to the different lines of investigation.

The following publications relating to bee culture, prepared in the Bureau of Entomology, are for free distribution and may be obtained by addressing the Secretary of Agriculture:[5]

[5]

Farmers' Bulletin No. 59, "Bee Keeping," and Farmers' Bulletin No. 397, "Bees," have been superseded by Farmers' Bulletin No. 447.

Circular No. 79, "The Brood Diseases of Bees," has been superseded by Farmers' Bulletin No. 442.

Bulletin No. 1, "The Honey Bee," has been discontinued.

Farmers' Bulletin No. 447, "Bees." By E. F. Phillips, Ph. D. 1911. 48 pp., 25 figs.

A general account of the management of bees.

Farmers' Bulletin No. 442, "The Treatment of Bee Diseases." By E. F. Phillips, Ph. D. 1911. 22 pp., 7 figs.

This publication gives briefly the symptoms of the various bee diseases, with directions for treatment.

Circular No. 94, "The Cause of American Foul Brood." By G. F. White, Ph. D. 1907. 4 pp.

This publication contains a brief account of the Investigations which demonstrated for the first time the cause of one of the brood diseases of bees, American foul brood.

Circular No. 138. "The Occurrence of Bee Diseases in the United States. (Preliminary Report.)" By E. F. Phillips, Ph. D. 1911. 25 pp.

A record of the localities from which samples of diseased brood were received prior to March 1, 1911.

Bulletin No. 55, "The Rearing of Queen Bees." By E. F. Phillips, Ph. D. 1905. 32 pp., 17 figs.

A general account of the methods used in queen rearing. Several methods are given, so that the bee keeper may choose those best suited to his individual needs.

Bulletin No. 70, "Report of the Meeting of Inspectors of Apiaries, San Antonio, Tex., November 12, 1906." 1907. 79 pp., 1 plate.

Contains a brief history of bee-disease investigations, an account of the relationship of bacteria to bee diseases, and a discussion of treatment by various Inspectors of apiaries and other practical bee keepers who are familiar with diseases of bees.

Bulletin No. 75, Part I, "Production and Care of Extracted Honey." By E. F. Phillips, Ph. D. "Methods of Honey Testing for Bee Keepers." By C. A. Browne, Ph. D. 1907. 18 pp.

The methods of producing extracted honey, with special reference to the care of honey after it is taken from the bees, so that its value may not be decreased by improper handling. The second portion of the publication gives some simple tests for adulteration.

Bulletin No. 75, Part II, "Wax Moths and American Foul Brood." By E. F. Phillips, Ph. D. 1907. Pp. 19-22, 3 plates.

An account of the behavior of the two species of wax moths on combs containing American foul brood, showing that moths do not destroy the disease-carrying scales.

Bulletin No. 75, Part III, "Bee Diseases in Massachusetts." By Burton N. Gates. 1908. Pp. 23-32, map.

An account of the distribution of the brood diseases of bees in the State, with brief directions for controlling them.

Bulletin No. 75, Part IV. "The Relation of the Etiology (Cause) of Bee Diseases to the Treatment." By G. F. White, Ph. D. 1908. Pp: 33-42.

The necessity for a knowledge of the cause of bee diseases before rational treatment is possible is pointed out. The present state of knowledge of the causes of disease is summarized.

Bulletin No. 75, Part V, "A Brief Survey of Hawaiian Bee Keeping." By E. F. Phillips, Ph. D. 1909. Pp. 43-58, 6 plates.

An account of the beekeeping methods used in a tropical country and a comparison with mainland conditions. Some new manipulations are recommended.

Bulletin No 75, Part VI, "The Status of Apiculture in the United States." By E. F. Phillips, Ph. D. 1909. Pp. 59-80.

A survey of present-day beekeeping in the United States, with suggestions as to the work yet to be done before apiculture will have reached its fullest development.

Bulletin No. 75, Part VII, "Bee Keeping in Massachusetts." By Burton N. Gates. 1909. Pp. 81-109, 2 figs.

An account of a detailed study of the apicultural conditions in Massachusetts. The object of this paper is to point out the actual conditions and needs of beekeeping in New England.

Bulletin No. 75, Contents and Index. 1911. Pp. vii+111-123.

Bulletin No. 75, Parts I-VII, complete with Contents and Index. 1911. Pp. viii+123.

Bulletin No. 98. "Historical Notes on the Causes of Bee Diseases." By E. F. Phillips, Ph. D., and G. F. White, Ph. D., M. D. (In press.)

A summary of the various investigations concerning the etiology (Cause) of bee diseases.

Technical Series, No. 14, "The Bacteria of the Apiary with Special Reference to Bee Diseases." By G. F. White, Ph. D. 1906. 50 pp.

A study of the bacteria present in both the healthy and the diseased colony, with special reference to the diseases of bees.

Technical Series No. 18, "The Anatomy of the Honey Bee." By R. E. Snodgrass. 1910. 162 pp., 57 figs.

An account of the structure of the bee, with technical terms omitted so far as possible. Practically all of the illustrations are new, and the various parts are interpreted according to the best usage in comparative anatomy of insects. A brief discussion of the physiology of the various organs is included .

BUREAU OF CHEMISTRY.

Bulletin No. 110, "Chemical Analysis and Composition of American Honeys." By C. A. Browne. Including "A Microscopical Study of Honey Pollen." By W. J. Young. 1908. 93 pp., 1 fig., 6 plates.

A comprehensive study of the chemical composition of American honeys. This publication is technical in nature and will perhaps be little used by practical bee keepers, but it is an important contribution to apicultural literature. By means of this work the detection of honey adulteration is much aided.

HAWAII AGRICULTURAL EXPERIMENTAL STATION.
HONOLULU, HAWAII.

Bulletin No. 17, "Hawaiian Honeys." By D. L. Van Dine and Alice R. Thompson. 1908. 21 pp., 1 plate.

A study of the source and composition of the honeys of Hawaii. The peculiar conditions found on these islands are dealt with.